I0479603

MATHEMATICS

THE ABSURDITY OF NUMBERS

BY

M.F. ONUCHUKWU

© 2019

**To
My Creator,
I give all the glory and
adoration for bestowing me
with the great insight used in
writing this book.**

Preface

The reader is expected to approach this book with great patience because some of the assertions made in the book may take him into unfamiliar territories.

The reader must also keep an open mind when reading the book because it will help him in assimilating the new ideas written in the book.

The reader must also do well to understand the definitions given for the various words used in explaining why the existence of numbers are absurd.

Contents.

Chapter

Substance and The Error of Trying to Quantify It – Chapter 1

The remarkable thing about substance is that it must exist (that is its essence is existence or "existing"). It is in the inherent nature of substance to exist. No other thing exists outside substance, No other thing can exist but substance. Substance is what causes all things to exist. Without substance nothing exists. Thus, you cannot argue that anything exists outside substance because doing so will make substance finite. Substance isn't finite; the reason being that substance has infinite things (modes of substance) which it causes into existence. The infinite nature of substance makes it impossible to have another substance. If there were to be another substance out there, substance immediately ceases to be infinite, which quite frankly in reality is impossible. Substance is also eternal meaning that it exists without having a beginning or an end. However, in this chapter, our focus shall be on the infinity of substance. The infinity of substance makes it impossible to divide it. Substance cannot be divided because it is infinite. The mode of substance is a state of substance which involves a particular feature of substance. The modes of substance are derived from their attributes; and are finite and temporal (has duration). They're conditioned to act in a definite manner. The error of quantifying substance has led to the introduction of numbers in Mathematics. The contemporary

Mathematician has failed to understand the infinity of substance and feels it is okay to quantify substance. The reality is that substance though having infinite attributes is really one and the same substance whose nature doesn't change. The modifications of substance don't change the nature of substance. Substance must always exist and its attributes must be infinite and eternal. Everything that exists must exist in substance. The modes of substance though they exist, don't contain the essence of substance which is existence. The attributes of substance are those things perceived by the human intellect as being the essence of substance. The important infinite quality possessed by substance (as a result of its existence by self-conception) causes infinite individual things (modes) into existence. By virtue of this causation, all modes of substance exist. The modes of substance don't have their essence as existence; this is true if you consider the temporal existence of the various modes of substance. It is not in the nature of the modes of substance to exist. It is the infinite nature of substance that makes it impossible to count or quantify or measure it. Substance cannot be counted or measured or quantified. The key question here is, what the need of numbers when in reality only substance actually exists. Numerology as we understand it today is very unnecessary especially if you considered that even the finite things have the nature of substance which in fact is the only thing that existed. The drudgery of counting or measuring or quantifying substance makes it a wild goose chase; the reason being that you cannot count or measure or quantify infinity. The falsities of numbers are understandable if and only if you considered that in reality only substance exists. Objects are counted or measured or quantified by the mathematician because his mind has been accustomed to seeing substance as finite and temporal or his mind has been accustomed to modality, whereas in the real sense of it, substance is infinite and eternal. Everything that exists is infinite because everything that exists is substance. Substance makes up the essence of everything that it conditions into existence and nothing can exist outside substance. Numbers in the reality cannot be used to label substance which is infinite in nature. Numerology has been the bane of Mathematics because it tries to quantify substance and it has continually failed to do so. In reality, numbers shouldn't exist because they can't accurately quantify substance. Isn't it sheer folly to try to quantify substance which is infinite? The symbols of numerology that we see today weren't in the real sense being used to quantify substance, rather, they were being used to describe the peculiar characteristics of the modes of substance which always exist in substance. The infinite nature of substance makes it impossible to measure it. In the same vein in which you cannot measure substance, you cannot also measure or quantify anything that exists in substance because they contain the infinite essence of substance. The various numbers being employed to count or measure or quantify things make no meaning to a mind that has been attuned to understanding substance as infinite. No one who understands the nature of substance can assert that there are so and so amount of substance. It is very ridiculous to assert such because substance cannot be counted or measured or quantified. One cannot say that there is "one substance" because

that will make substance finite. By

extension, one cannot count or measure

or quantify the modes of substance

which contain the infinite essence of

substance. If everything that exists

(including humans) is substance, then

why try to count or measure or quantify

it? Whatever exists exists because of

substance. All modes of substance exist

because substance exists and they

contain in them the infinite essence of

substance which makes it absurd to

quantify the things that exists because of substance. The failure of the mathematics of numbers lies in the fact that it fails to account for the infinity of substance. According to modern day mathematics, we tend to count or measure or quantify objects according to certain criteria such as their similarities. The idea behind the creation of numbers was to count or measure or quantify objects which we encounter in our everyday lives. Objects despite being modes of substance are very much substance. They are substance because their nature expresses a particular (conditioned) finite and temporal quality of substance. Thus, you can't count or measure or quantify them. For example, if you sought to measure the weight of a steel ball, the error in doing so lies in the fact that you'll end up measuring the effect of an attribute of substance on a mode of substance, which sincerely is an absurdity. All modifications of the attributes of substance may be counted or measured or quantified in the modal sense but not in reality. Since all modifications exist in and are conceived through substance, they've the infinite and eternal nature of substance which makes it impossible to count or measure or quantify them.

The Falsity of Zero – Chapter 2

The reality of our existence lies in the fact that zero doesn't exist. It never existed and it doesn't exist. It is a negation to truth to admit that zero exists. If we're to go back to the definition of substance, we shall quickly realize that substance conceived itself and exists through itself. Substance as a result of its essence always existed. There was nothing before substance, substance is all that existed. There was no point in time when the world was without form or void. There never was because substance always existed. There are no two substances (it is absurd to assert that there are two substances in existence because it will make substance finite). The human intellect perceives that substance has as one of its essence, infinity. When something is infinite, it's wrong to assert that such a thing can be nothing. In the real sense of it, zero in mathematics would mean that something doesn't exist and not the absence of substance. Modern mathematics through the symbol of zero has asserted that substance doesn't exist which is very absurd. If the mathematicians had asserted that the modality of substance could be zero then they'll be right. For them to assert that zero is the absence of substance is in itself very erroneous. If zero means the absence of substance, then should one mean the presence of substance? The error of mathematics lies in the fact that it's trying to quantify what's infinite and not quantifiable as a result. The error of mathematics lies in the fact that it's trying to deny the existence of substance. It forgets that all the modifications of substance are derived from the attributes of substance which are infinite and eternal. By default, one will expect the modes of substance to have the nature of substance which is infinity and eternity. To be very honest with the reader, zero shouldn't be treated as a number having its own right whatsoever because in the first place, it is a falsity. Zero is a falsity because it's "undefined". Have you ever tried dividing zero by zero, it gives you an error because even the computer understands that zero doesn't exist. One can't assert that there isn't anything when in reality there is substance. One can't also assert that there wasn't anything before substance because that statement denies the very essence of substance which is existence. Zero can't express lack of quantity either because there shouldn't be any concept like "quantity" in the first place. One can't validate the absurdity of trying to quantify infinity by asserting that there is a lack of substance when in reality, everything which exists, exists in and through substance. One can't validate zero either by establishing that there could be an absence of substance. The greatest absurdity can be seen when some people erroneously assert that adding zero to "positive" numbers yields "positive" numbers whereas adding zero to "negative" numbers yield "negative" numbers. The first question to be asked because of this ridiculous assertion is this, in reality, is there any such thing as zero? In reality can you quantify substance with numbers? If you can't quantify substance with numbers, why should numbers be employed in the counting or measurement or quantification of substance? What do they mean by assigning negatives and positives to numerical symbols? These are a few questions that ought to be answered speedily by those who argue that zero according to the definition they give exists. From the perspective of reality, it is an anomaly to compare zero with void, simply because substance is all that exists even in what appears to be empty space. Empty space couldn't exist without substance. Nothing exists without substance (no mode exists in itself; modes always exists in substance). Thus, the concept of empty space being represented with zero is also erroneous and unfounded. Vacuum which you may refer as an "empty space" has also been erroneously been represented with "zero". Vacuum they assert is empty of all "substances". How can vacuum be empty of "substance"? The reader understands that by now, it is impossible for vacuum to exist outside of substance. Since substance is all that exists, vacuum couldn't be empty of substance because that will mean vacuum exists outside of substance which is very absurd. The truth is that vacuum exists because of substance and as such exists in substance. To represent vacuum with zero (which they assert to be nothing) is wrong because it's impossible to have nothing when there is substance. Mathematicians and philosophers because of the doubts they cast on zero ponder till date on the question, "how can nothing be something?" It is very sad to see them ponder on what shouldn't be pondered on in the first place and what actually has a straight forward answer. The answer to their question is very simple. There isn't anything like "nothing" in the first place. But there was always "something" which we call "substance". One can assert that there question is very rhetorical because substance can never be "nothing". The reader must be careful to differentiate between a mode of substance not existing and the assertion that nothing exists. If one asserted that the rivers can laugh, you must realize how untrue that is, because a river cannot laugh. This is very different from saying that substance doesn't exist which will be very absurd. No matter the symbols used to represent zero, it doesn't take away the fact that asserting that in reality "nothing" exists is untrue and very erroneous. The reality is this, substance exists and thus something exists. Nothing exists outside substance. The proper use of zero will involve two uses: (1) to represent that nothing can exist besides substance (2) to represent things that couldn't exist in reality. To employ zero to represent the nothingness of substance is in reality false.

The Meaningless Number One – Chapter 3

The crux of this chapter lies mainly in exposing how meaningless it is to assert that there's "one" substance. It was clearly stated in the previous chapters that in reality only substance exists. It was also asserted that substance cannot be divided at all because substance is infinite and eternal. It was also asserted that everything which exists is the sum total of substance and the modes of substance. It was also asserted that the

modes of substance couldn't exist
outside of substance and they only exist
in substance. It was also asserted that in
reality the modes of substance contain
the infinite and eternal nature of
substance despite being finite and
temporal from a modal perspective. It
was also asserted that the modes of
substance are derived from the attributes
of substance which is basically the
human intellect perceives the essence of
substance to be. Since substance must
exist by it's nature, it must exist to cause

other things into existence. That is to say, it must exist in infinite ways and all of this infinite ways of causing existence (the attributes of substance) is unknown to man. The human intellect can perceive only the thought and movement attributes of substance. The human intellect also understands that there can't be "one" substance because quantifying substance means asserting that there is another substance out there. It is absurd to assert such. However, one must be careful when describing the number one, because in reality it doesn't exist since you can't have one substance. You only have substance and the modes of substance (which is basically still substance but in different states of existence). For emphasis sake, having one substance makes substance finite and in that case it ceases to be substance. The reader must understand by now that it is impossible for substance to be finite. It is also absurd to claim that substance is "single" when in reality substance cannot be measured. It is also absurd to say that substance is "single" and made up of many parts when in reality substance cannot be divided. It is impossible to divide substance because substance is indivisible. The infinite nature of substance makes it impossible to measure or divide it. The error of mathematics lies in the fact that it views substance from the modal perspective and not from the point of reality. The problem with viewing substance from the modal angle is that in doing so, one fails to consider the essence of substance.

By doing so, one chases the shadow of substance instead of substance itself. The mode of substance are derived from the attributes of substance and are merely the particular types or forms of substance conditioned to necessarily take on such a form by substance. But they are still substance. This is the gist of this book. The fact that particular types of substance are in existence doesn't change them from being substance. They're still substance regardless. The key question to be asked is this, isn't it very absurd to assert that after zero comes one? That is according to those who assert such, that after the "absence of substance comes "one" substance". The statement sounds like something out of a horror movie to one who understands substance. It doesn't make any sense to those who understand that substance cannot be counted or measured ot quantified because it is infinite. By now, the reader must have also understood that multiplication in reality is meaningless. In modality, it could be useful if you considered the forms of substance but in reality, it is absolutely useless to multiply substance because doing so will mean that substance can be divided or that substance can be more than one which in reality is very absurd. Also worthy of note is the fact that it's impossible to say that there's "one" substance.

Nevertheless, there arises a situation where "one" can accurately be used without ambiguity such as: (1) to describe the uniformity of substance (2) to describe the indivisibility of substance. The author believes that this was what the ancients had in mind when they called forth into their thoughts the symbol that represents "one". The error in Mathematics lies in the fact that the numeric values they use most often than not tend to fall within the interval of zero to one which really has no meaning. In other words, the numerical values they employ tend to fall within the absence of substance and "finite" substance. It's very clear that the numeric values employed in Mathematics and normalized to fall within the intervals of the so called "zero" and "one" is done erroneously from a realistic perspective. Another definition of the number one which is false is that one is the probability of an event that is almost certain to occur. The falsity of this definition lies in the fact that there's nothing probable about substance. Substance must always exist and this isn't a probability but a certainty. Thus, it is very wrong for one to assert that probability governs substance. This is the real error of Mathematical probability and in reality doesn't portray the fact that certainty governs substance. For something to occur in reality, it has been conditioned by substance to happen that way. The modes of substance are produced from the attributes of substance which causes their existence. Thus, one can assert that they are the effects produced by the fact that substance exists. The effects produced by substance are always conditioned. This conditioning makes one understand that substance operates with absolute certainty. Thus substance behaves in a certain way and not in an "almost certain way". For example standing under the rain without a covering from the rain will get one wet. This is a certainty and definitely, there's no probability in that assertion. You can't represent such effects produced by rain as "one" because it is almost certain to happen. By representing such occurrence with "one" it shows that the assertion is derived purely out of a lack of understanding of reality. It also shows that the individual doesn't understand that "one" as a number is impossible and full of absurdity. The usage of "one" in everyday life to quantify substance shows to a large extent the level to which the human mind has misunderstood substance and its nature. Also "one" cannot be used to describe the ultimate reality because substance which is the ultimate reality isn't "one" but infinite. To assert that substance is "one" removes infinity and eternity from substance; such an assertion is very absurd. We've been able to prove that "one" as a number is meaningless, thus the reader must also conclude that other numbers invented by the Mathematicians must also be meaningless. Such conclusion is correct if you consider the impossibility of quantifying substance and its modes.

The Nonsensical Natural Numbers – Chapter 4

Nature doesn't involve numbers.

Numbers can't be said to fall within the

sphere of nature. Numbers were

invented by humans. Substance doesn't

involve numbers because substance

can't be counted or measured or

quantified. In mathematics, natural

numbers are defined as numbers used

for counting and ordering. In previous

Chapters, the author proved

convincingly that reality doesn't involve

numbers. The author also proved that substance is infinite and thus cannot be divided into parts and that the division of substance will suggest the existence of numbers which in reality is very absurd. However, the author must reiterate the need for the reader to be able to distinguish reality from modality. Modalities involve particular forms or types of substance and are distinguishable from one another. The error that arises from thinking modal wise is that modality doesn't take into recognition the fact that only substance exists and that even the modes of substance contain the essence of substance (called attributes when perceived by the human intellect) which makes them substance. Modality also deals mainly with the imagination. Mathematicians have always imagined that numbers exist because of the different modifications of substance in existence. Part of the definition of natural numbers involves "ordering". Substance cannot be ordered because ordering substance will mean that substance contains parts which need to be ordered. To assert that substance has parts is to assert that substance is finite and such an assertion is very absurd. Substance is nature itself because everything which exists exists in substance. There can't be anything in existence which exists outside substance. Thus, it is unreasonable to assert that there can be anything like "natural substance" because it will mean that substance exists in nature which is such a nonsensical assertion. If substance is the same as nature, it is illogical to also have the words "natural substance" in existence because such words will connote a tautology. Natural numbers are said to to begin with either zero or one. The author was able to prove in Chapters two and three that the existence of "zero" was false and that "one" as a number was meaningless. Mathematicians have asserted that the existence of natural numbers was a direct consequence of the human "psyche". The conjuration of numbers into existence was borne purely out of the erroneous belief that substance can be divided into parts and hence can be counted or measured or quantified. The human "psyche" or mind can perceive the essence of substance just as it can actively imagine that the modes of substance (which are still substance) can be divided into parts. A lack of understanding of the nature of substance is as a result of the imagination which comes from a passive mind. An active mind seeks to understand the essence of substance and as a result doesn't fall into the errors associated with a passive mind. Thus, it is very clear to the reader that the existence of natural numbers is as a result of the human mind dwelling in a passive state and harboring such false ideas which they readily imagine to exist. In mathematics, natural numbers are also defined as a class of all sets in one to one correspondence with a particular set. The usage of the word "set" in defining natural numbers proves that the existence of "natural numbers" is indeed absurd. The reason being that "set" involves being finite. A set is a group of similar things that belong together in some way. From a modal perspective, sets are modes of substance. A mode of substance is a particular type or form of substance which in reality can have other similar types or forms in existence which it shares common characteristics or properties with each other. The reader at this point, understands that modes of substance are finite because their

essence doesn't pertain to "existing".
They are also finite because they've
other particular types or forms like them
in existence. One will expect sets to be
finite as modes of substance are finite.
Such an expectation won't be wrong.
Sets are finite and as such don't involve
the infinite essence of substance. In
reality modes of substance cannot be
counted or measured or quantified
because they contain the infinite essence
of substance and are as a result still
considered substance. It has also been

asserted by mathematicians that the set of all natural numbers is an infinite set. The absurdity of this definition lies in the fact that it is senseless to assert that sets or modes of substances can be infinite. In short it's impossible to have an infinite set because sets aren't infinite but finite. Infinity only pertains to substance but not to the modes of substance. Thus, it's risible to assert that sets are infinite. To even assert the existence of "countable infinity" is the greatest absurdity ever. How can you count infinity? How can you count what's limitless? How can you count what's boundless? Whatsoever is said to be "infinite", cannot be counted or measured or quantified. It's also impossible for one to assert that substance can be added to substance because that'll mean that there exists more than one substance, thus making substance finite which in reality is very absurd. For the mathematician to remotely suggest that substance can be divided or multiplied shows that he fails to understand the very nature of substance. If substance can be divided or multiplied, it'll mean the existence of more than one substance or the existence of the parts of substance which in reality is very absurd. To think in the modal sense is to exclude reality which involves the essence of substance. It's to assume that substance can be counted or measured or quantified. Substance can't be said to be associative or commutative or distributive either. If one asserted that substance is associative, it'll imply that substance has parts and such an assertion is very absurd. Also, if substance was said to be commutative, it'll mean that substance can be counted or measured or quantified which in reality is very farcical. Finally, if substance was asserted to be distributive, it will also mean that substance can be divided and this assertion is indeed very absurd.

The Risibility Of Prime Numbers – Chapter 5

In mathematics, prime numbers are known to be natural numbers and in reality, they don't exist. In this book, the author dedicated a full chapter explaining the nonsensical natural numbers and the absurdities surrounding their false existence therein. In that particular chapter, it was shown that numbers don't fall within the premise of nature or substance. It was also proven that the existence of numbers were entirely farcical since it was impossible to divide substance. According to Wikipedia, a prime number is a natural number greater than "one" that cannot be formed by multiplying two smaller natural numbers. From the definition of prime numbers, it's very clear to the reader that prime numbers are conjured from the false idea that there can be more than "one" substance which in reality is very impossible. Substance is infinite and as such cannot be divided into parts or said to be "one" for that will imply that substance is finite, which is very absurd. In reality, "primality" doesn't pertain to substance because substance isn't divisible. Substance only exists or is "existing". To assert that there are infinitely many prime numbers is also wrong. It's like asserting that there are infinitely many substances which have been divided into parts. It's impossible to assert that substance is finite and at the same time infinite. Such contradictions are very absurd. Another false conjecture made by the mathematicians is that there are infinitely many pairs of prime numbers having just one even number between them. This conjecture brings one back to the fact that it's very contradictory to ascribe infinity to finite things. The author has reiterated throughout the book that it's absurd for numbers to exist let alone prime numbers. If the mathematicians imagined that numbers existed because of his lack of understanding of substance, then it'll be okay if we concluded that the mathematician imagined that substance can be divided into parts. The key question to be asked at this point is this, how is it that the mathematician asserts that what can be divided into parts is infinite? Substance on its own can't exist in pairs because such an existence will make substance finite which in reality is impossible. To also assert that one can construct a perfect number is indeed very senseless. Perfection can only be associated with substance in so far as it's said to have everything

necessary for existence. Substance is the

cause of existence and nothing can exist

outside substance. Thus, because

substance has everything necessary for

existence, we say that it's perfect. The

error of mathematics lies in the fact that

numerical symbols are used to count or

measure or quantify substance. In reality

if one took out time to understand

numerical symbols as employed by the

ancients, they'll come to the conclusions

that the numerical symbols weren't

employed in the quantifying of substance. Rather they were employed to cement the fact that different modes of substance were finite and can be represented as numerical symbols. The use of numerical symbols also displayed the solemn fact that modes of substances were inter-related and existed only in substance. Cryptography as an art clearly lends credence to the fact that numbers can only be used to "represent" the different modes of substance and not used to count or measure or quantify substance. To remotely suggest that numbers can be even or odd clearly shows that the mathematicians doesn't understand that numbers are an absurdity. Substance can't be divided or leave a remainder when it's supposedly divided. The impossibility of even and odd numbers lies in the fact that substance is infinite and as such is indivisible. For the mathematician to assert that prime numbers are the basic building blocks of natural numbers, verifies the fact that he has no basic understanding of substance and its modes. Another erroneous claim about prime numbers is that the sequence of prime numbers never ends. In mathematics, numbers are employed to do the impossible, which is the counting or measurement or quantifying of substance. If the nonsensical existences of numbers falsely suggest that substance is finite and temporal, how then can the mathematician now conclude that the same numbers can never end, since it was established that substance is the only thing that exists?

For the umpteenth time, substance is all

that exists. Substance is infinite and

eternal. Substance can never be counted

or measured or quantified as well as it

can never end (that is it enjoys infinite

existing). In reality numbers don't exist

because substance can't be counted or

measured or quantified. Also in reality,

using numerical symbols to represent the

different modes of substance in

existence shows that the modes of

substance are finite. In the modal sense,

a sequence or progression of numerical symbols is finite and points to the fact that the different modes of substance which they are used to represent are finite. It has been suggested that prime numbers indicate "minimality" or "indecomposability". Substance has nothing to do with the "barest minimum" because substance cannot be counted or measured or quantified. Substance also can't be divided up to the point where division was no longer possible because that'll mean that substance is finite and this is very impossible. The bane of quantum mechanics lies in the fact that it proposes that substance can exist in discrete quantities. It is impossible for substance to exist in small quantities because it will mean that substance can be counted or measured or quantified, which is very absurd. Finally, substance cannot be factored because it will suggest that substance has parts which in reality aren't obtainable.

The Impossibility of Negative Numbers – Chapter 6

According to Mathematics, numbers can be negative and if they aren't negative, then they are positive. The impossibility of having negative numbers lies in the fact that nothing can be less than substance. The definition of negative numbers clearly shows that the mathematician intended to prove that there were numbers less than zero. In chapter two, the author was able to prove that zero is a falsity and as such shouldn't exist because there can never be an absence of substance. If the absence of substance was proven to be an absurdity, the key question to be answered here is this, is it possible to have anything less than the "absence of substance"? The answer to the question is very rhetorical. There can never be anything less than substance because less than substance will make substance become finite which in reality is very absurd. The two main problems with asserting that negative numbers exists is this: (1) there isn't anything like the absence of substance because substance must always exist. (2) there isn't anything less than substance because substance is the only thing that exists. To assert that there's anything less than substance means that there is another substance out there with a lesser quantity. This assertion quite frankly makes substance finite and is definitely absurd. Substance is self-conceived and thus caused itself to exist. This fact makes negative numbers have no legitimacy because it is obvious that substance can never ever be less than itself. The modes of substance don't make substance change its nature or essence. The modes of substance are the only particular types of substance and as such don't in anyway make substance lesser in quantity because substance can never be counted or measured or quantified as shown in Chapter One. The error of mathematics in using negative numbers to "solve problems" extends to the quadratic equations which really are an effort to represent substance as a set of finite numbers and is totally absurd. Substance can't be counted or measured or quantified, so it's a waste of precious time to trying to do otherwise. The error of mathematics in using negative numbers also extends to accounting where negative numbers are used in reality to represent debts. This is very absurd if you consider that all which exists is substance. It's impossible to quantify substance or to have "less than zero substance because it will mean that there's an absence of substance and that also substance is finite. Thus, representing debt using negative number is absurd because debt is a mode of substance and as well substance. The result of the preceding assertion in reality is that it's erroneous to employ the word "negative numbers" because it certainly makes no sense to do so. The irony of referring negative numbers as "absurd numbers" is that the usage of numbers is also very absurd considering that everything which exists is substance and cannot be counted or measured or quantified because of the fact that substance is infinite and eternal. The impossibility of the negative numbers also lies in the fact that they were employed to exclude the evidence of substance. By now, the reader must understand that it's very impossible to exclude substance. How do you exclude the only thing which exists? It is necessary for the reader to understand that substance is what causes existence and as such cannot be excluded from substance. Everything exists in substance and employing "numbers" to denote that for some reasons yet unknown that there's a possibility of substance momentarily disappearing is very absurd. The only negation of substance that exists is when the mind conjures up thoughts which are very impossible. For example it is a negation to say that a stone speaks Latin because going by the conditioning of stone, it is impossible for stone to speak let alone speak Latin. The thought of such an idea is in itself a negation of truth and very well doesn't fall within the premise of using numbers to count or measure or quantify substance. The author must state without mincing words that negative statements exist and they are statements that negate reality. The reader must be able to differentiate this from "negative numbers" which truly are absurdities. Substance as a result of its primary essence must exist. The negation of substance is impossible no matter the angle from which you think about it. Substance cannot have an anti-substance or what the scientist terms as "anti-matter" because that will make substance two and thus finite, which is absurd because substance shall always remain infinite. It's very obvious that nothing like anti-substance exists because it will mean that substance no longer caused itself to exist but depends totally on another thing for it's existence.

This will make substance finite and temporal which is absurd.

The Ridiculousness Of Rational and Irrational Numbers - Chapter 7

According to the mathematicians, rational numbers can be expressed as the quotient of two integers whereas irrational numbers cannot be expressed as the quotient of two integers. In Mathematics, an integer can be defined as a number that can be written without a fractional component. These definitions are given so that the reader may perfectly understand the error of employing numbers to count or measure or quantify substance. Also, the reader at this point is very conversant with the fact that substance cannot be divided because it is infinite. An integer is a number and as a number, its existence is very absurd because numbers shouldn't have been invented in the first place. Numbers shouldn't have been used to count or measure or quantify substance because it is impossible to count or quantify or measure substance. It is quite obvious that rational numbers involve decimals and the existence of decimals from a reality viewpoint is very absurd. It is impossible to say that substance can be a fraction or be divided. Substance cannot be divided because it's infinite. Rational numbers can always be expressed as a ratio of two integers. What this means is that in reality, mathematicians presume that numbers can actually be used to count or measure or quantify substance and as a result believe that certain numbers when expressed as quotients yield results which they believe is rational. Substance can't be divided into parts, this is a fact which you can't wish away. To express substance as a quotient is in reality impossible and mathematicians have failed to recognize that it's an error to do so. One can't assert that substance is rational because substance has nothing to do with rationality. Substance only has to do with existence. However, if you considered substance and the attributes of thoughts that it has, it isn't still safe to assert that substance is rational because rationality has no dealings with the infinite essence of substance. It is purely within the premise of modality to discuss about the rationality of the mode of substance. Substance thinks thoughts and so does humans (a mode of substance). However, substance thinks infinite thoughts while humans due to the fact that there are other humans in existence (particular types or forms like them) think only finite thoughts. However, this doesn't make the human thoughts different from the thoughts of substance since human thoughts are contained in substance. Now, it is essentially absurd to conclude that substance is rational because human thoughts can be channeled in such a way that they think along the lines of reason and self-preservation. This conclusion is strictly within the lines of modality and not reality. In reality, substance thinks thoughts which are absolutely infinite. It is absurd to give the attributes of human thoughts to that of substance. Rationality and irrationality cannot be said to be the characteristic of substance. The fact that the human thought exist in substance doesn't make substance finite. Since substance is purely infinite, it makes perfect sense to state that rationality or irrationality cannot be said to belong to the purely infinite substance. The irrationality of numbers doesn't make sense as well. Numbers have no place in reality and as such attributing irrationality to the so called "numbers" is certainly very wrong. If one took out time to understand the role of decimals in determining whether numbers can be described as rational or irrational, one will come to a very quick conclusions that decimals are in reality the bane of mathematics. For the umpteenth time, it is impossible to divide substance because substance is infinite. One cannot divide what is deemed infinite into parts because doing so, will stop such a thing from being infinite. This is the real crux of the matter that modern mathematicians have failed to

understand. The assertion that substance

cannot be divided into parts is an eternal

truth which the human intellect doesn't

fail to appreciate when it clearly

understands the assertion. This chapter

has emphasized that rationality and

irrationality can only be ascribed to the

human mind and not substance. The

human mind is able to understand that

substance cannot be deemed rational or

irrational because substance is infinite.

Even if you considered the attributes of

the thought of substance, one will

clearly understand that substance still

has infinite attributes of thought because

there isn't any other substance out there

that shares the attributes of thought with

substance. The mode of substance (the

human mind in this case) can be

irrational if it doesn't possess the power

of understanding the emotions. However,

if the human mind is trained to

understand the emotions, it shall tow the

lines of reason and self-preservation.

This isn't the case with rational and

irrational numbers which in reality are absurdities. History has it that Pythagoras believed in the "absoluteness of numbers" and never accepted the existence of irrational numbers. It's very clear to the reader that Pythagoras's belief in the "absoluteness of numbers" makes no sense if one considered the fact that numbers are being used in a wrong way, which is to count or measure or quantify substance. What exactly does Pythagoras mean by the "absoluteness of quantifying substance"?

One with a perfect understanding of substance can see that Pythagoras has no similar understanding of substance, hence the assertion of his preference for the absoluteness of "numbers". What Pythagoras failed to understand is that only substance can be absolute. Only substance exists. Nothing else exists but substance and nothing can ever exist but substance. By calling forth numbers to count or quantify or measure substance, Pythagoras showed that he didn't understand the nature of substance when he resorted to using numbers to quantify substance. The fact that the so called "higher degree equations" were developed and that they could only be solved using irrational numbers showed the absurdities that were prevalent in mathematics in the 19th century. It was on the background of these errors that modern mathematics set its foundations.

The Preposterous Transcendental and Real Numbers – Chapter 8

The human intellect despite being finite is able to perceive the infinite essence of substance (that is the attributes of substance). In Mathematics, the transcendental numbers are those numbers whose existence they say is beyond the limits of human knowledge. The reader must understand that numbers don't fit into the premise of reality because reality entails nothing but substance. It is an absurdity when numbers are used to count or measure or quantify substance. Substance is infinite and cannot be counted or measured or quantified. It is greater absurdity when numbers are said to be "transcendental". That is to say that there are certain numbers beyond the human intellect. The truth is that the existence of numbers is based on the false idea that substance is divisible or finite. And to say that there is certain knowledge beyond the human intellect is wrong also. The human mind which is a mode of thought is finite and always understands the nature of substance. The human mind in existence shall always have knowledge of substance. The knowledge of substance is the highest knowledge the human mind can possess. However, to have this knowledge of substance by the human mind is possible if you considered the fact that the human mind exists in substance and derives its characteristics from the attribute of thought of substance. Let's assume that the mathematicians understood substance and also theorized that the knowledge of substance was beyond the human intellect, they will also be very wrong. To assert that the knowledge of substance is beyond the human intellect is to say that the human mind doesn't exist in substance. Such an assertion is very absurd. The highest knowledge the human mind can have is that of substance and this knowledge is not beyond the human mind. Transcendental numbers exists in the realm of mathematics because mathematicians have failed to understand the very nature of substance which is the same as the essence of substance. According to mathematicians, real numbers on the other hand are defined as values of continuous quantity that can represent a distance along a line. The problem with the definition of real numbers lies in the fact that "quantity" is a farce. Quantity itself doesn't exist because substance can't be counted or measured or quantified. Since quantity doesn't exist, how can one argue that quantity is continuous? Continuity can never be applied to quantity because quantity is a total farce. Continuity can't be applied to substance because it'll mean that substance stopped its existence at one point and continued its existence at another point. Substance is eternal because it always exists. There's nothing like duration with substance. Substance has no beginning or end. Substance always exists and as such one can't ascribe time or duration to substance. Only eternity can be applied to substance because substance is self-conceived and caused other things into existence (modes). Modes of substance are temporal because they've for their existence other particular types or forms which have their existence in substance. Substance is eternal because the human intellect perceives substance as that which exists. If you take away "infinite existing" from substance, it ceases to be substance and it becomes temporal.

Substance is also considered eternal because its existence can't be counted or measured or quantified. That is to say it enjoys infinite "existing". It's in the premise of substance to exist and as such nothing else exists but substance (that's why substance's existing is infinite). It isn't in the mode of substance to exist though they (that is the mode of substance) have in reality the essence of substance. In modality, modes of substance have duration if you consider the fact that they're finite (that is they have other particular types or forms like them in existence). The human mind has the essence of substance because despite being a mode of substance, it's still substance. For the umpteenth time, modes of substance are in essence substance because they exist in substance and express a particular attribute of substance. In the case of the human mind, the attribute of substance being expressed is thought. When you consider the fact that substance thinks, one shall quickly come to the conclusion that the human mind contains eternally an essence of substance which is thought. That's why the human mind is eternal. Continuity can't be associated with substance because continuity suggests duration and quite frankly, substance cannot have a duration. Substance enjoys infinite "existing" because in reality, nothing else can be said to be existing but substance. Real numbers are also said to have the ability to represent a distance along a line. The reader must understand that the word "distance" is in reality also a farce. It's also one of the erroneous conjectures made in mathematics in the effort to count or measure or quantify substance. A line in reality doesn't exist. What exists is substance and the attribute of substance that pertains to movement. The effort to try to "measure distances" is in fact a mental aberration on the part of the mathematician and a display of his lack of understanding of substance. The modern mathematician lives in a modal world which in all honesty isn't the world of reality. The modal world is a

world of shadows and abstraction. It is

the world of imagination and not a world

of concrete facts. In Chapter Five, the

author clearly showed how the concepts

of rational and irrational numbers were

ludicrous. If we defined real numbers as

including both rational and irrational

numbers, it will automatically mean that

the concept of real numbers is absurd

based solely on the facts established in

Chapter Five. Transcendental numbers

which are included within irrational

numbers are definitely not left out in the fact that their existence represent absurdities. Modern mathematics also asserts that the set of all real numbers is uncountable. This assertion is very wrong because numbers aren't suppose to be in existence. Only substance is uncountable, that is to say only substance is infinite. Substance is infinite because of the fact that only substance exists and everything else exists in substance.

The Farcical Infinity In Mathematics – Chapter 9

A popular thought in Mathematics was that one can divide infinity into parts. They believed that one can take a part from infinity and add a part to infinity. The mathematicians didn't know that the only thing which can be infinite is that whose caused itself into existence and whose essence pertains to "existing". Substance exists. Nothing exists but substance. The fact that only substance exists makes it infinite. It's impossible to divide anything that's infinite. According to the mathematicians, infinity can be distinguished into actual infinity and potential infinity. What exactly they mean by actual infinity and potential infinity is yet to be understood. How can infinity be deemed to be actual when the nature of infinity involves indivisibility and not actuality? To assert that substance is actually substance is to commit tautology. It is to assert that there's a huge possibility that substance may not be actual, which is very absurd.

Infinity can't be said to be potential

because the nature of infinity doesn't

involve potentiality but indivisibility.

Substance can't be deemed to be

potential because substance can never

change it nature or stop existing. There

was never a time when substance didn't

exist and thus had a potential to exist,

neither was there a time when substance

existed and thus had the potential to stop

existing. Infinity doesn't involve

existence or the likelihood of existence.

Infinity involves only indivisibility. Mathematicians also assert that mathematics is the science of the infinite. The key question raised by this definition of mathematics is this, what does the mathematician mean by the infinite? By infinite, does he mean substance? Or Does he mean numbers? If mathematics involves the science of substance, is it safe to say that thus far, the mathematicians aren't dealing with substance? The knowledge of substance isn't a science at all because the human mind exists in substance to know substance. The thoughts that involve the knowledge of substance are the noblest which man can attain. A man who attains such is blessed. However, care must be taken not to posit this knowledge as a science when indeed it isn't one. Scientific knowledge is obtained through a rather rigorous process of observation, hypothesis, experimentation and inferences but the understanding of substance can be acquired by purely thinking about the nature of God. If mathematics was the science of numbers, then it's safe to assert that the very foundation of mathematics is illogical. Numbers as the reader must have come to understand is in reality very unreasonable. Numbers are very absurd because they're employed in doing the impossible. Numbers are employed in the counting or measurement or quantifying of substance which is very absurd. Numbers have their place only when we describe modality and not reality.

Numbers should've no place in reality because they fail to describe the essence of reality which is existence. They undermine substance and its essence. Numbers treat the modes of substance like they're not substance. The absurdity of numbers cannot be overemphasized. In mathematics, it's asserted that infinity can be used to denote an unbounded limit. The nature of infinity doesn't involve limitation. But one can assert that infinity means limitless. Anything which is limitless cannot be divided. To divide a limitless thing means that such a thing is no longer limitless but limited. The absurdity of defining infinity as "unbounded" limit shows that the mathematician doesn't understand the nature of limitation. How can what is said to have limit be "unbounded"? The word limit connotes boundaries. It is impossible to have "unbounded" boundaries. It makes no sense to assert that what has boundaries can be unbounded. It isn't in the nature of substance to have limits. Substance is always limitless because it is the only thing that exists. Substance is all that exists. Nothing else exists but substance. All other things exist because of substance and they exist in substance. Numbers aren't infinite. In reality numbers represent the greatest absurdities. Substance is the infinity that the mathematicians speculate with numbers. Substance is infinite because the nature of substance has to do with "existing" or existence. Anything whose nature pertains to "existing" or "existence" is infinite and eternal. The human mind is eternal because it possesses the eternal essence of substance. The human mind through the intellect is able to perceive its eternal nature. The human mind is only considered a mode of substance because it's united to the human body. That's is when one considers the human body. However, if the human mind was to be considered as being a mode of thought whose existence is in substance, then one shall understand that the human mind has the essence of eternity found in substance. Mathematicians have also pondered on the nature and size of the observable universe. The key question they asked about the universe is this: is the universe finite or infinite? Before we proceed to answer that key question, it is important that universe is defined. According to Wikipedia, the universe can be defined as all of space and time and their contents including planets, stars, galaxies, and all other forms of matter and energy. Going by this definition of the universe, it is safe to conclude that the universe is finite because of the time factor mentioned in the definition. Time or duration can only be attributed to modes of substance which are finite. Substance enjoys infinite "existing" because nothing else exists but substance. There's substance and substance alone. It's in the nature of substance to exist and nothing else can exist but substance. Also, one cannot count or measure or quantify substance and its infinite essences and eternity is an essence of substance. Thus duration can't be applied to substance. The definition of the universe will be wrong If we considered the fact that space was included in it. According to Wikipedia, space is the boundless three-dimensional extent in which objects and events have relative position and direction. The error in this definition lies in the fact that it's very absurd to put together "boundless" and "three-dimensional" in the same definition. Only substance can be boundless or limitless or infinite.

Whatever is boundless or limitless or infinite cannot be said to be "three dimensional". That is, it cannot be said to be counted or measured or quantified. This is where the definition of space is wrong because the mathematician makes an error when he asserts that "three-dimensional" is boundless. If we asserted that the "universe" contained all forms of matter and energy, does that make the universe a substance? The answer is 'NO'. The universe from the definition given has a duration since it involves time. It's rather correct to posit that the universe is a mode of substance. It's also clear that in this case (that is from the definition of the universe), one can assert that the forms of matter and energy are also modes of substance since they all exist in substance. Same goes for the planets, stars and galaxies that make up the given definition of universe. The fact that the universe is a mode of substance from the given definition means that there are other particular types or forms like the universe in existence. The reason for

making this assertion, the author

believes is very clear to the reader.

The Ludicrous Imaginary and Complex Numbers – Chapter 10

According to Wikipedia, an imaginary number can be defined as a complex number that can be written as a real number multiplied by the imaginary unit. Also, according to Wikipedia a complex number can be defined as a number of the form a + bi, where "a" and "b" are real numbers and "i" is an indeterminate. In Chapter Six, real numbers were discussed and it was proven that their existence was preposterous. The author will discuss this Chapter from the viewpoint that real numbers form the foundation of imaginary and complex numbers. The existence of real numbers were described as being ludicrous because (1) they are numbers and ought not to exist because substance which they seek to count or measure or quantify cannot be counted or measured or quantified. (2) they involve "distance" which seeks to count or measure or quantify the infinite attribute of movement (3) they involve lines which insinuate that substance can be counted or measured or quantified. The imaginary unit of imaginary numbers confirms the fact that numbers are derived from the imagination and not from reality. All false ideas are derived from the imagination because they don't seek to describe reality. The imagination can be said to be derived from the human mind working in modality. A man may imagine that a hill dances but in reality that isn't the case. If and only if the man understood the nature of hills,

he would know that it was impossible

for a hill to dance by its conditioning.

Thus, it's safe to assert that the

imagination gives the mathematician the

false idea that substance can be divided

into parts (the existence of numbers).

When the human mind receives

impressions from a body, it imagines

that body to be present even after that

body is removed. The imagination of

numbers is not different because after

seeing numerical symbols, the mind

always imagines such numerical symbols as present. It associates the images of these numbers with one another without understanding the very nature of numerical symbols employed. This is exactly why numbers are in usage today; the mathematicians have operated in the realms of imagination instead of reality. From the definition of complex numbers, it is clear that it existence is absurd because the foundations of its definitions are based on real numbers and imaginary numbers.

It's also very important that we mention the fact the "i" in the definition of complex numbers is said to be "indeterminate". When considered from the viewpoint of substance, it is also very clear to the human intellect that indeterminate doesn't pertain to substance. Substance cannot be said to possess indeterminate characteristics. Everything in existence is determined and conditioned to exist in a particular form (modes of substance). Substance causes everything to exist through a determinate and conditioned chain series of causes and effects. Thus a mode of substance A is caused into existence by another mode of substance B and B is caused into existence by mode of substance C to infinity. This shows that everything in existence has for the cause of its existence substance. If substance is the first cause of existence, it follows logically that due to its infinite nature that there will a chain series of causes starting from substance to infinity. In Mathematics, it's also asserted that complex numbers can be equal to each other if and only if both their real and imaginary parts are equal. The reader must understand that substance has nothing to do with equality but everything to do with existence. If one asserted that "equality" pertains to substance, it will mean that substance can be divided into parts, which practically is impossible. Thus "equality" doesn't fall within the premise of substance. In Mathematics, the nature of complex numbers involves the square roots of negative numbers. The reader must remember that negative numbers don't exist in reality and that also it's impossible for negative substance to exist in reality. In Chapter Four, the author clearly asserted that negative numbers were an impossibility in reality. Negative numbers will mean that there was something less than substance and that substance was finite as a result of the existence of this very thing which was less than substance. Such an assertion is very absurd. Square roots involve a number multiplying itself to produce a particular number. If one considered square roots from the perspective of substance, it will be very clear that substance cannot be replicate itself, let alone double itself for that will mean that substance is finite and such an assertion is very absurd.

Conclusions

In a world full of false ideas, it's easy for one to be misled by such ideas. In reality, substance is all there is in existence. Existence or "existing" pertains only to substance. The world has been conditioned to believe that substance can be counted or measured or quantified. This erroneous belief has led to the use of numbers in the daily lives of men. Men have been led to buy the idea that numbers can be used to count or measure or quantify substance. This idea is false and is borne out of the imagination of men. It's an idea developed from passive minds who think that substance can be counted or measured or quantified. Substance has as one of its essence, existence. The earlier people embrace this basic truth, the faster the world realigns to reality. If substance is that there is, it makes sense to say that substance is infinite and eternal. Substance is infinite because only substance exists. Substance is eternal because it enjoys infinite "existing". Nothing else can be said to possess eternity but substance. Substance also has attributes which basically is what the human intellect perceives the essence of substance to be. Apart from being infinite and eternal, substance is also said to think and move. The ancients must have had a different understanding of substance and numbers. Numbers to them were mere representations of the different modes of substance and revealed the inter-relatedness of the different modes of substance. They firmly understood that substance cannot be counted or measured or quantified and didn't bother trying to do so. Some mural paintings by the ancient Egyptians reveals this truth clearly. However, some people whom the author has discussed with have been very concerned about the fate of the global monetary system if numbers aren't to be used in counting or measuring or quantifying value. The reality here is that the global monetary system isn't actually governed by the values given to the digits placed on the

paper currency. Now, before we proceed,

the reader must understand that false

wise, the value of numbers are always

represented with symbols (numerical).

These symbols as the author has earlier

explained can also be used to represent

the different modes of substance. Paper

currency is a mode of substance

(particular type or form of substance).

One can represent the paper currency

with any symbol he wishes to. In

modern times, digits (numerical symbols)

are written on paper currencies. It's easy to be misled that these digits represent a counted or measured or quantified value of substance. In reality, the digits don't represent substance but the mode of substance which in this case is the paper currency. Also in reality, the numerical symbols imprinted on the paper currency only show the inter-relatedness (how modes of substance relate) between the paper currency and the other different modes of substance in existence. The global monetary system will not be threatened at all if the numerical digits imprinted on the paper currency were to be replaced with any symbol that's not the numeric symbol. The symbols are only there to mark the paper currency which is the mode of substance that has a relationship with other modes of substance. This explanation on the marking of paper currencies with numeric symbols will help the reader understand how numerical symbols are to be employed in everyday use. It will further throw light on the fact that substance can't be divided and as such

eternally infinite.

THE END.

www.ingramcontent.com/pod-product-compliance
Lightning Source LLC
Chambersburg PA
CBHW030652220526
45463CB00005B/1750